PREFACE

Over the years, many lives have gone, houses/offices and cars worth millions of dollars destroyed in fire incidents. A good look at this seems as if this ugly situation is not ending soon.

Fire outbreak, as we all know cannot ordinarily start up on itself except in a rear occasions. This means each and every known fire outbreak started from a source which ordinarily would have been prevented.

In the light this **"CAUSES AND PREVENTION OF HOUSE, OFFICE AND CAR FIRE"** was born. This book is meant to educate the world populace the causes and prevention of house, office and car fire with the aim of eradicating/eliminating incidences of fire outbreak in our homes, offices and cars.

It is worthy to note that most fires are as a result of lack of awareness on damning consequences and some of our attitudes that enhance chances of fire outbreak.

For better understanding and base on the context of this book fire is defined as the rapid oxidation of a material in the exothermic chemical process of combustion releasing heat, light, and various reaction products. The flame is the visible portion of the fire. The burning flame is typically yellow or orange and there is smoke. If there is not enough oxygen available during a chemical reaction, incomplete combustion occurs, and products such as carbon (C) and carbon monoxide (CO) as well as water and carbon dioxide are produced.

Fire in its most common form can result in conflagration, which has the potential to cause physical damage through burning. The negative effects of fire include threat to life and property, atmospheric pollution, and water contamination. Fires start when a flammable or a combustible material, in combination with a sufficient quantity of an oxidizer such as oxygen gas or another oxygen-rich compound, is exposed to a source of heat and fuel. Fire cannot exist without all of these elements in place and in the right proportions.

Once ignited, a chain reaction must take place whereby fires can sustain their own heat by the further release of heat energy in the process of combustion and may propagate, provided there is a continuous supply of an oxidizer and fuel.

A structure (house, office or car) fire is a fire involving the structural components of various types of residential, commercial, industrial buildings or cars. Residential buildings range from family homes, townhouses or various commercial buildings ranging from offices to shopping malls.

With our most valued possessions at risk, the old adage "prepare for the worst and hope for the best" is a good approach. Taking precautions to prevent the most common causes of house, office and car fire is imperative. Most of the time, common sense can prevent house, office or car fire. However, distractions could occur and accidents can happen at any given moment in homes. The best way to prevent house/office fires is to always remain alert and keep an eye on your heat sources.

Fire can be extinguished by removing any one of the elements of the fire triangle.

Many fatal fires start at night, when only a working smoke alarm can save your life. Install one alarm per level of your house, particularly near the kitchen and sleeping areas. Replace the batteries as and when due, and test your alarm monthly. Make sure everyone in your house and office recognizes the sound of the alarm and knows the nearest exits. And do not forget to practice your home or office fire escape plan regularly.

House or car fires can be devastating. The worst of them can leave an entire family homeless or loss of property and feeling as though they have lost everything. Thankfully, there are steps that can be taken to prevent or at least decrease house, office and car fire.

Within the context of this book, we would consider possible causes and prevention of house, office and car fire. This book is intended to teach the world populace possible causes of house, office and car fire and how best to prevent it from occurring. Here are different types of situations and conditions that can cause fire outbreak and also the best possible means to prevent it.

TABLE OF CONTENTS

CHAPTER ONE

CAUSES AND PREVENTION

OF

HOUSE AND OFFICE

FIRE

1.1 COOKING FIRES

The major cause of house fires is cooking. Open flames from the stove and intense heat in the oven easily result in a fire when unsupervised. Most often food or cooking tools catch fire and quickly lose control. While giving up cooking all together may be a bit extreme, paying extra attention and never leaving the kitchen while preparing food is an easy way to reduce this risk.

Pots and pans can overheat and cause a fire very easily if the person cooking gets distracted and leaves cooking unattended. Always stay in the kitchen, or ask someone to watch your food, when cooking.

You may think it is okay to step outside and talk to a neighbour, take out the trash, check the mail or do any other activities while you have food cooking on the stove. However, all it takes is one splash of hot oil on a hot burner to start a fire. Leaving food cooking unattended could be a recipe for disaster, so always stay close to the kitchen to prevent a fire outbreak.

Surprisingly, kitchens provided for staff are often at the source of fires in offices. Always ensure that staff do not leave food unattended while it is cooking. This rule will help reduce the risk of a fire starting in the office.

It is important to fit kitchens or other cooking areas with fire blankets, extinguishers and automatic fire detection. All electrical appliances in the kitchen should be regularly tested by a qualified electrician and confirm them fit for use. Be on alert! If you are sleepy or have consumed alcohol do not use the stove or stovetop. Stay in the kitchen while you are frying, boiling, grilling, or broiling food. If you want to leave the kitchen for even a short period of time, turn off the stove. If you are simmering, baking, or roasting food, check it regularly, remain in the home while food is cooking, and use a timer to remind you that you are cooking. Keep anything that can catch fire e.g. oven mitts, wooden utensils, food packaging, towels or curtains away from your stovetop. Do not leave anything unattended. Turn off the heat if you need to leave the room or house to help your kids, take a phone call, or answer the door. Do not overfill pots and pans with oil. Do roll up long sleeves and tie back long hair. Keep cloths, kitchen curtains, towels, pot holders, and paper towels away from the stove. Keep your appliances clean.

This is a big one! Pay particular attention to your oven and stovetop: food residue and oil build up is flammable. Do have a fire extinguisher on hand. Installing a fire extinguisher in your kitchen is not a bad idea either; If you have a small cooking fire (from the cooking pot) and decide to fight the fire; on the stovetop, smoulder the flames by sliding a lid over the pan (or pot) and turning off the burner. Leave the pan covered until it is completely cooled. For an oven fire, turn off the heat and keep the door closed.

Cooking is the leading cause of house fires and home-fire injuries, with the kitchen topping the list as the most dangerous room in your house when it comes to fires.

Let us be on guard!

Another method to help deal with a potential fire is to have an interior anti-fire sprinkler system installed, especially in areas that are higher risk. This can help limit the damage if a fire broke out, and prevent any fires from spreading to different parts of the building.

If you ever have an oil fire, never put water on it.

1.2 CHRISTMAS TREE FIRES

Every winter/ Christmas season, families all around the world bring large trees into their homes to celebrate a beloved holiday. However, Christmas trees can easily become a significant fire hazard. Dry trees combined with strings of lights quickly turn into fire sticks. Before placing the tree in your home cut off an inch or so from the bottom of the trunk, removing any dead wood that would prevent the tree from soaking up water. Be vigilant about watering your tree, keep it hydrated! Also be sure to turn off Christmas tree lights when you go to sleep at night or leaving the house. Or avoid the risk altogether by getting an artificial tree!

Keep the tree in a stand that will hold two to three litres of water, and top it up daily. Keep the tree away from all heat sources, including radiators, furnace ducts, television sets and fireplaces. Check decorative lights before placing them on the tree, and discard any frayed or damaged lights/cords. Never place candles on or near the Christmas tree.

From the moment a spark ignites a Christmas tree, it can take only a matter of minutes before noxious smoke and scorching heat fill the entire room, igniting everything within it.

Christmas trees are powerful fuel sources, especially when dry. The tree becomes a fuel that burns very rapidly and gives off a lot of heat energy. A blazing tree can give off as much heat capable to create flashover within a room, where all objects begin burning virtually at the same time.

The heat gets trapped at the ceiling, then banks down, forming a hot gas layer full of smoke. This combination of heat and smoke makes it impossible for anyone in the room to survive and could make it difficult for those in other parts of the house to escape. The rapid nature of Christmas tree fires makes escaping a fire difficult.

For that reason, taking precautions when choosing and decorating a tree becomes extremely important.

1. **Give live Christmas trees a fresh cut.** Always choose a freshly cut Christmas tree so it will absorb water and stay fresher longer. "Sap flows out of trees, so without a fresh cut at the bottom, there might not be up-take of water.
2. **Water your tree daily.** Constant watering keeps trees fresher longer, but the moment the tree appears to drop its needles, it is a sign that it is drying out.

3. **Use approved lights and connect them properly.** Choose an approved and tested lights and avoid connecting multiple extension cords.
4. **Inspect lights and decorations.** Before decorating your tree, lay out strings of lights and look for any broken or missing lights. Needles can get stuck in empty light sockets, creating a potential fire hazard. Electric energy passes through the bulb sockets and can cause needles to ignite.
5. **Toss damaged lights and decorations.** Do not attempt to repair light strings if they are worn, frayed or show other problems. Throw them away and buy a new set of lights.
6. **Choose your tree's location carefully.** Place the tree away from stairs, where fire can quickly travel to bedrooms. Avoid placing it near heat sources, such as a wood, stove or fireplace. Being close to radiators and heat vents can make the tree quickly dry up.
7. **Avoid using candles near the tree.** Candle itself is a measure source of fire outbreak; but keeping it close to Christmas tree makes it more dangerous. Do not keep candle close to Christmas tree.
8. **Avoid combustible ornaments.** Pinecones and other ornaments can add fuel to a Christmas tree fire and should be avoided.
9. **Keep pets safe.** Pets can chew, paw and otherwise damage lights and potentially knock over the tree.
10. **Unplug at night.** Never leave the tree plugged in when you are away from home or asleep.
11. **Close bedrooms doors.** Closing your bedroom doors at night can keep out harmful smoke and flames in the

event of a Christmas tree fire, giving you more time to escape.
12. **Test smoke alarms.** Make sure smoke alarms are properly located and in good working condition.

1.3 ARSON/INTENTIONAL FIRES

Arson is the crime of wilfully and maliciously setting fire to or charring property. The term arson can also refer to the intentional burning of other things, such as motor vehicles, watercraft, or forests. The crime is typically classified as a felony, with instances involving a greater degree of risk to human life or property carrying a stricter penalty. A common motive for arson most times is to commit insurance fraud. In such cases, a person destroys their own property by burning it and then lies about the cause in order to collect against their insurance policy.

A person who commits arson is called an 'arsonist'. Arsonists normally use an accelerant (such as gasoline, kerosene) to ignite, propel fires.

The Prevention: This might seem obvious, but there are ways to decrease the Arson or intentional fires:

- Know your neighbours, and watch out for each other's properties, especially during any vacation or absence.

- Report unoccupied/unused buildings in your area to city officials and ensure that they have doors, windows, and other openings secured.

- Clean up, dispose of, or otherwise remove any potential targets or materials that could be used to fuel the fire: piles of leaves, garbage, old furniture that's been left outdoors, abandoned vehicles, gasoline and other flammable liquids.

Also, use close circuit television cameras and motion sensor lights. Suitable security system will make arsonist think twice.

1.4 SMOKING FIRE

Smoking is the source of many of house fires. Lit cigarettes accidentally dropped on any household items can quickly become large fires if not timely managed. Only smoking outside the house will reduce this risk significantly or otherwise designate a portion around your house as a smoking designated area.

Bedrooms are best to be kept off limits for smoking. A cigarette that is not put out properly can cause a flame, as the butt may stay alit for a few hours. It could burst into flames if it come into contact with flammable materials, such as furniture, clothes or any other combustible materials.

After a late night party or stressful day, it may be tempting to have stick of cigarette just before bed. With the possibility of dozing off, there is a tendency to toss the butt anywhere while it has half-lit. If that lands on the rug or mattress, the consequences can only be imagined. Smoking is the leading cause of home fire deaths – let that sink in. If you do smoke, make sure that your cigarette/cigar/pipe is fully extinguished, and never leave them unattended. It is common for some products to overbalance as they burn down. Be sure that your ashtray is heavy enough to not tip, made of non-flammable material, and verify that the contents of your ashtray are cold before emptying it into the garbage, preferably outside the house. Other tips on smoking fire include:

- It is safer to smoke outside, but make sure cigarettes are put right out and disposed of properly.
- Never smoke in bed, and avoid smoking on arm chairs and sofas – especially if you think you might fall asleep.
- Take extra care when you are tired, taking prescription drugs or if you have been drinking alcohol.
- Use proper ashtrays, which cannot tip over and stub cigarettes out properly.
- Do not balance cigars or cigarettes on the edge of an ashtray, or anything else – they can tip and fall as they burn away and cause a fire.
- Do not leave lit pipes or cigarettes unattended.
- Always empty ashtrays carefully. Make sure smoking materials are out, cold and preferably wet them before throwing into a bin – never use a wastepaper basket.

- Keep matches and lighters out of children's reach, and buy child resistant lighters.
- Never smoke if you use healthcare equipment like medical oxygen or an air flow pressure relief mattress. If you use paraffin-based emollient creams, ask for non-flammable alternatives instead.
- Consider additional safety measures such as fire retardant bedding or nightwear.

Since smoking has no health benefits, it is advisable you quit!

1.5 FIRE DUE TO HEATING-RELATED EQUIPMENT/APPLIANCES
Unplug small appliances when they are not in use examples: toasters, hair straighteners, Pressing iron, kettles, etc. Keep appliances clean, well-maintained, and away from heat areas. Do not continue to use any malfunctioning appliances; get them fixed or replaced. Keep clothes, curtains, and any other potentially combustible items a minimum of three feet from all heating devices.

Dryers: make sure it is installed and serviced by a professional, and do not forget to clean out the lint filter every time. All venting materials should be rigid or flexible, and the air exhaust vent pipe should be free of restriction with the outdoor vent flap opening during the dryer's operation. Once a year, or more if required, clean lint out of the vent pipe or get a professional to do it for you.

If you own a wood-burning fireplace or stove, ensure that it is properly installed, well maintained, and do not forget to open the damper, and keep it open until the fire is out and the ashes are cool enough to touch.

Dishwashers: Have you ever opened your dishwasher immediately after it finished its cycle? The heat that shoots out when you open the door can be overwhelming. A dishwasher contains heating elements that dry your dishes. These heating elements get wet, heat up, and cool down every time you use the dishwasher. Old or faulty elements can start a fire. Never turn on a dishwasher before leaving your home.

Microwaves: Microwaves are very convenient. Warming up leftover meals can be a big timesaver when the family schedule is hectic. They also can be very dangerous. There have been many complaints about microwaves starting on their own and causing a fire. If you hear your microwave running, and you didn't turn it on, unplug it immediately.

Refrigerators: One might never think of a refrigerator being a fire risk; however, an overheated compressor can cause fire. In addition, a light that stays on all the time can be hazardous.

Toasters: Toasters have electric elements inside that are used to brown your bread or bagel. If the toaster fails and doesn't turn off, a fire can start. Never leave a toaster unsupervised and remove the crumbs on the bottom of the toaster regularly. In addition, inspect the electric elements for stuck crumbs.

1.6 CANDLE FIRE

Candles turn to liquid in order to release their fragrance and this wax carries heat and that can cause another surface, such as a wood table, to catch fire. Over half of candle fires start because the candle is too close to combustible materials.

Most of the incidents are due to negligence. Leaving a lit candle unattended can quickly lead to a fire. Numerous flammable materials in the vicinity of a candle can easily catch fire. Reduce this risk by trimming the wick of your candle to help limit the size of the flame, or use battery operated candles.

Candles look and smell pretty, but if left unattended they can cause a room to easily burst into flames. Keep candles away from any obviously flammable items such as books and tissue boxes. Always blow a candle out before leaving a room.

We put candles and leave them beside curtains or clothes or even on tables made from wood. The candle may fall off or a nearby object may get burned. Regardless of how relaxing candles may be, do not leave them unattended, and be sure they are out before falling asleep or leaving the room, etc. Watch out for curtains, and ensure that your candles are out of reach of any children and pets to ensure that they cannot be bumped over.

A lit candle in an unoccupied room is just asking for trouble. Candles are a major cause of home fires, especially during the winter, harmattan and Christmas holidays, where they are often placed too close to blankets or holiday decoration.

Prevention:
- **Keep candles away from children and pets:** While candles look nice on the coffee table and add ambiance to your living room, a passing child or wagging tail can easily tip them over.
- **Keep matches and lighters in a safe place:** Lighting materials should be stored up high and out of sight in a kitchen cupboard, pantry, or even a closet.
- **Avoid using candles in the bedroom:** Approximately one-third of candle fires start in a bedroom and one-half of fire deaths happen between midnight and 6:00 am.

- **Have several candle holders on hand:** Candles come in many shapes and sizes so you should have the right-size holder for the candles you like to burn. And it is important to make sure candle holders are placed on durable heat-resistant surfaces.
- **Toss the candle if it is two inches or less:** Replacing a candle more frequently costs far less than replacing your home. Do not let a candle burn to close to the holder.
- **Avoid using water to extinguish the candle:** Hot wax can splatter in all directions if doused with water. And the temperature change could cause a glass container to crack or break. Consider using a snuffer to extinguish the candle.
- **Never use candles during a power outage or as a night light:** Flashlights, or other battery powered lights, are much safer than candles. Night lights are also much safer and fairly inexpensive.
- **Always keep an eye on the candle:** Do not leave a candle in an unattended room for a significant period of time.
- **Always follow the manufacturer's safety recommendations:** Manufacturers want you to enjoy their candles so follow their recommendations.
- **Use common sense:** Make sure the area around your candle is free of clutter. Also watch for sporadic airflow around the candle which can cause the flame to shift direction. And always make sure the candle is at least 12 inches away from other household items that can burn.

1.7 ELECTRICAL EQUIPMENT FIRE

An electrical appliance, such as a toaster can start a fire if it is faulty or has a frayed cord. A power point that is overloaded with double adapter plugs can cause a fire from an overuse of electricity. A power point extension cord can also be a fire hazard if not used appropriately.

Negligence and "do not care attitude" is one of the main causes of electrical fires in an office environment. Typically, faulty wiring, not servicing or not taking general care of your equipment is a common causes of fire due to electrical equipment.

Never overload lead extension socket by plugging in appliances that together exceed the maximum current rating stated for the extension lead. This could cause a plug terminated in an electrical wall socket to overheat and possibly cause a fire.

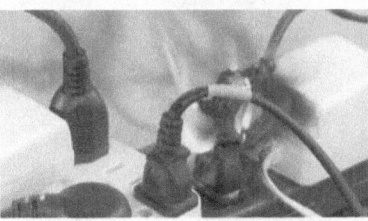

Taking precautions like not stacking or covering electrical hardware that requires air to circulate around it will help avoid any risk of the equipment overheating and catching fire.

Lamp shades and light fittings can build up heat if they are very close to light globes. Lamp bases can become a hazard if they are able to be knocked over easily, and so should be removed if they are. Check that down lights are insulated from wood panelling or ceiling timbers.

The common causes of electrical fires include

1. **Faulty outlets & appliances.** Most electrical fires are caused by faulty electrical outlets and old, outdated appliances. Other fires are started by faults in appliance cords, receptacles and switches. Never use an appliance with a worn or frayed cord which can send heat onto combustible surfaces like floors, curtains, and rugs that can start a fire.

    Running cords under rugs is another cause of electrical fires. Removing the grounding plug from a cord so it can be used in a two-prong electrical outlet can also cause a fire. The reason appliances have the extra prong is so they can be only used in outlets that can handle the extra amount of electricity that these appliances draw.

2. **Light fixtures.** Light fixtures, lamps and light bulbs are another common reason for electrical fires. Installing a bulb with a wattage that is too high for the lamps and

light fixtures is a leading cause of electrical fires. Always check the maximum recommended bulb wattage on any lighting fixture or lamp and never go over the recommended amount.

Another cause of fire is placing materials like cloth or paper over a lampshade. The material heats up and ignites, causing a fire. Faulty lamps and light fixtures also frequently result in fires.

3. **Extension cords.** Misuse of extension cords is another electrical fire cause. Appliances should be plugged directly into outlet and not plugged into an extension cord for any length of time. Only use extension cords as a temporary measure. If you do not have the appropriate type of outlets for your appliances, hire an electrician to install new ones.

4. **Space heaters.** Because these types of heaters are portable, many times people put them too close to combustible surfaces such as curtains, beds, clothing, chairs, couches and rugs. Coil space heaters are especially dangerous in this regard because the coils become so hot they will almost instantaneously ignite any nearby flammable surface.

   If you do use space heaters, use the radiator-type that diffuse heat over the entire surface of the appliance. These are less likely to ignite flammable items, but should still be kept away from them.

5. **Wiring.** Outdated wiring often causes electrical fires. If a home is over 20 years old, it may not have the wiring capacity to handle the increased amounts of electrical appliances in today's average home, such as computers, wide-screen televisions, video and gaming players, microwaves and air conditioners.
   Breakers should be triggered when circuits get overloaded by too much electricity, but outdated breaker boxes often have worn connectors that do not work, causing the system to overload and start an electrical fire.

Double check the appliances and power points in your home and office and ensure the following:

- Your electrical appliances do not have loose or frayed cords/plugs
- Your outlets are not overloaded with plugs
- Do not run electrical wires under rugs or heavy furniture.

Electrical fires are one of the most common threats homeowners face. Warning signs include fuses that blow or circuit breakers that trip frequently, or lights that dim when you use another appliance.

1.8 CURIOUSITY FIRE

Children cause fires out of curiosity (what happens when something burns) or mischief (they're angry, upset or destructive, and fire is a major taboo to break). Kids can cause a fire out of curiosity, to see what would happen if they set fire to an object. Keep any matches or lighters out of reach of children, to avoid any curiosity turned disaster.

Install a smoke alarm in your child's room and practice a home escape plan with your children and family in case there was a fire. Kids may be involved in fire play if they find matches or lighters in their room/possession. If you smell sulphur in their room, and/or find toys or other personal effects that appear melted/singed signalled that indeed they are playing matches/lighters.

1.9 BARBEQUES FIRE

Barbeques are great for an outdoor meal, but should always be used away from the home, tablecloths or any plants and tree branches. Keep barbeques regularly maintained and cleaned with soapy water and clean any removable parts.

There is nothing like outdoor grilling. It is one of the most popular ways to cook food. But, a grill placed too close to anything that can burn is a fire hazard. They can be very hot, causing burn injuries. Follow these simple tips and you will be on the way to safe grilling.

Check the gas tank, hose for leaks before using it for the first time each year. Apply a light soap and water solution to the hose. A propane leak will release bubbles. If your grill has a gas leak, by smell or the soapy bubble test, and there is no flame, turn off both the gas tank and the grill. If the leak stops, get the grill serviced by a professional before using it again.

Please note:

- Propane and charcoal barbeques should only be used outdoors.
- The grill should be placed well away from the home, deck railings and out from under eaves and overhanging branches.
- Keep children and pets at least three feet away from the grill area.
- Keep your grill clean by removing grease or fat build-up from the grills and in trays below the grill.
- Never leave your grill unattended.
- Always make sure your gas grill lid is open before lighting it.
- If you use a starter fluid, use only charcoal starter fluid. Never add charcoal fluid or any other flammable liquids to the fire.
- Keep charcoal fluid out of the reach of children and away from heat sources.
- There are also electric charcoal starters, which do not use fire. Be sure to use an extension cord for outdoor use.

- When you are finished grilling, let the coals completely cool before disposing in a metal container

1.10 FLAMMABLE/COMBUSTIBLE LIQUID FIRE

Flammable and combustible liquids are liquids that can burn. They are classified, or grouped, as either flammable or combustible by their flashpoints. Generally speaking, flammable liquids will ignite (catch on fire) and burn easily at normal working temperatures. Combustible liquids have the ability to burn at temperatures that are usually above working temperatures.

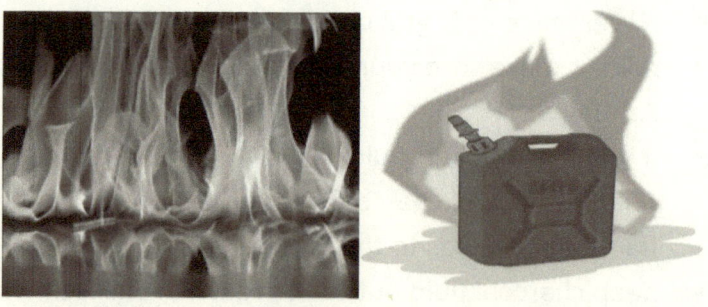

Flammable and combustible liquids are present in almost every workplace. Fuels and many common products like solvents, thinners, cleaners, adhesives, paints, waxes, kerosene, petrol, diesel and polishes may be flammable or combustible liquids. Everyone who works with these liquids must be aware of their hazards and how to work safely with them.

At normal room temperatures, flammable liquids can give off enough vapour to form burnable mixtures with air. As a result, they can be a serious fire hazard. Flammable liquid fires burn very fast. They also give off a lot of heat and often clouds of thick, black, toxic smoke.

Combustible liquids at temperatures above their flashpoint also release enough vapour to form burnable mixtures with air. Hot combustible liquids can be as serious a fire hazard as flammable liquids.

Spray mists of flammable and combustible liquids in air may burn at any temperature if an ignition source is present. The vapours of flammable and combustible liquids are usually invisible. They can be hard to detect unless special instruments are used.

Most flammable and combustible liquids flow easily. A small spill can cover a large area of workbench or floor. Burning liquids can flow under doors, down stairs and even into neighbouring buildings, spreading fire widely. Materials like wood, cardboard and cloth can easily absorb flammable and combustible liquids. Even after a spill has been cleaned up, a dangerous amount of liquid could still remain in surrounding materials or clothing, giving off hazardous vapours.

Vapours can flow from open liquid containers. The vapours from nearly all flammable and combustible liquids are heavier than air. If ventilation is inadequate, these vapours can settle and collect in low areas like sumps, sewers, pits, trenches and basements. The vapour trail can spread far from the liquid. If this vapour trail contacts an ignition source, the fire produced can flash back (or travel back) to the liquid. Flashback and fire can happen even if the liquid giving off the vapour and the ignition source are hundreds of feet or several floors apart.

If you have any flammable liquids in the home or garage such as petrol, kerosene or methylated spirits, keep them away from heat sources and check the label before storing. Be careful when pouring these liquids.

Be careful not use adulterated kerosene. Purchase your kerosene from Government approved filling station. Do not purchase your Kerosene from the unapproved vendor. Unexpected and deadly explosion could occur when using adulterated kerosene.

Storing area for kerosene and methylated spirited should be restricted to only authorized persons.

Petrol is never permitted to be stored at home because of its high flammability if not properly handled.

1.11 HOUSE OR OFFICE FIRE FROM BUSHFIRE

Many families have been rendered homeless as a result of bushfire. Bushfire gets to a building from various ways:

- **Gutters:** Overhanging trees can cause compacted leaf litter to build up in gutters. During a bushfire flying embers land in this material, catch alight and spread flames to combustible parts of the roof structure such as wooden facia boards, rafters, roof battens, and eaves. It is a good idea to clear out your gutters each year as part of seasonal bushfire preparation. Some people choose to wait until a bushfire is approaching to do this, but going onto your roof for the first time in semi-darkness while embers are flying at you can put you at risk, and endanger your life.

- **Vents and weep holes:** Together, vents and weep holes allow for fresh air to pass through a building and for excess moisture to leave, reduce condensation and mould. They are necessary for our comfort and health, and maintaining the integrity of a building. However in a bushfire these types of external openings can allow flying embers to enter the building and start spot fires. Having steel or other non-combustible mesh with small holes in front or behind vents and weep holes can reduce the bushfire risk while still allowing air and moisture to pass through.

- **Conduction & Convention:** Houses can also be caught up on bushfire by convention and conduction of heat. Bushfire either directly or indirectly will transfer heat to the building and in the presence of oxygen and fuel, all the materials in the building will build their required ignition temperature and they would be ignited.

Finally, create a buffer zone around your house, which should serve as fire break.

1.12 HOUESHOLD GENERATOR FIRE

In the recent times, many lives have been lost and properties destroyed as a result of fueling household generator when the engine is running and use of hand phones while refueling. Also, using generator without plug spark arrestor is a risk that jeopardize the livelihood of entire household if petrol comes in contact with plug terminal. The use of phone within an explosive and flammable atmoshere is suicidal; therefore, do not use your phones when fueling your generators.

Store fuel in a well ventilated area and away from ignition sources. Never refuel a running generator. Shut down your generator and fuel after about 5 minutes. If the gasoline or diesel fuel for your generator has sat for over 6 months, drain and replace with fresh fuel

Every year, people die from running their generators in their garage or too close to their house. Do not run your generator in your garage, even with the door open. And do not run it under your eaves either.

Generator fuel tanks are always on top of the engine so they can "gravity-feed" petrol to the carburettor. That setup can quickly turn into a disaster if you spill petrol when refuelling a hot generator. Think about it; if you spill fresh petrol onto a hot engine and it ignites. Spilling is easy if you refill especially at night without a flashlight. We know you can go without power for a measly 15 minutes, so cool your heels while the sucker cools down.

1.13 FIRE DUE TO OIL BEING TOO HOT

Heating up your oil too quickly will make it more prone to catching fire. Cooking with oil or grease is a primary cause of kitchen fires because the temperature of the oil quickly increases and can go from smoking oil to a blazing fire in no time.

If you think your house smells bad after the oil starts smoking and you have set off the smoke detector, just imagine how it would smell if the oil had caught fire.

1.14 FIRE DUE TO NOT BEING ALERT

Cooking late at night or early in the morning before you are completely alert is never good. Whenever you are working with an open flame, cooking with oil, or using a microwave or oven be sure you are alert and will be able to stay awake.

Many kitchen fires have been caused due to the individual cooking falling asleep. Be sure not to be distracted either by GSM calls or another errand that could take you out of the kitchen. Give undivided attention to the cooking and avoid any distraction. Such distraction could be that condition that could break the thin line and fire could start.

1.15 FIRE DUE TO UTENSILS TOO CLOSE TO HEAT SOURCE

Many home chefs have to multitask when cooking a meal. It is easy to set a cooking necessity such as an oven mitt near a heat source without realizing it.

Be aware of where you set everything and always make sure any utensils, paper towels or other items are far enough away from heat sources.

1.16 FIRE DUE TO NOT KNOWING HOW TO STOP A FIRE

A decision everyone has to make when a fire starts in the kitchen is "Should I attempt to put this fire out myself?" or call for help.

Household member should know what to do in the time of emergency. They should know how to use firefighting equipment (fire extinguisher, fire blanket) and all the escape routes in the house

1.17 STORAGE AND MATERIAL HANDLING FIRE

Pay considerable attention to the safe storage of flammable materials such as paper. It is vital to keep heat sources away from them. (Always take extra care when positioning any heat source in your office e.g. heaters).

Always follow the Safety Data sheet for instructions on material storage of all flammable liquids, glues and solvents as they are liable to combust unless stored and used properly.

1.18 GAS LEAK FIRE

A strong smell of sulphur or rotten eggs is a common sign of a gas leak. Leak points can omit a hissing or whistling sound as the pressurised gas is pushed through a small tear or loose connection. Larger leaks can produce a roaring sound from the gas line. A properly working stove burner burns with a crisp, blue flame.

If you can smell gas in the air, there is no need to panic. The volume of gas in the surrounding air needs to be very high for a person to suffocate or for it to reach ignition point.

But it is important to recognise that any gas leak has the potential to be dangerous if left unattended. Large quantities of gas, particularly in enclosed spaces, can result in unconsciousness or death caused by carbon monoxide poisoning, fire hazards and exposure to toxic by-products. Not all gas leaks are large enough to be detected by smell. It's important to regularly test your restaurant equipment. An easy way to check for a gas leak is using a bubble test.

Bubble tests are simply performed with a mixture of soap and water.

Spray the soapy solution around the pipes, fittings and valves or the gas bottle connection. The solution will bubble at the site of any leak.

If you find a Gas Leak:
- Turn off the gas supply at the stove or bottles.
- Open windows to allow extra ventilation and dispersal of the gas.
- If the leak seems large/very strong smell-evacuate family members from the building, especially if the area is small and confined.
- Do not do anything that could generate a spark.
- Do not switch off and/or on electrical sockets/switches except it is intrinsically safe
- Call your plumber & gasfitter to diagnose and repair the fault.

Do not use phones in the kitchen. In this era of phone and social media, many people go into the kitchen with their phones. The temptations are many. For one, an incoming call may provide a distraction too costly.

1.19 LIGHTNING/ THUNDERSTORM FIRE

The biggest danger lightning poses to a house is fire. Wood and other flammable building materials can easily be ignited anywhere an exposed lightning channel comes in contact with (or passes through) them.

Lightning passing through a house will often 'branch' and utilize more than one path to ground at a time. It can also jump through the air from one conductive path to another in what is called a *side flash*. For example, lightning may first connect to electric lines in the attic of a house, then jump to better-grounded water pipes on the first or second floor. Lightning can connect to gutters, then jump to a window frame as a 'stepping stone' to the electrical system or water pipes. All or part of bolts have been seen jumping from wall outlets to sink faucets and even across rooms!

Lightning current will produce significant damage to a house that is not equipped with a good protection system. Professionally installed lightning protection systems are expensive and the risk of a direct strike is low, so most homes do not have them.

Lightning presents three main hazards to a house that is hit directly:

- **Fire danger**: The biggest danger lightning poses to a house is *fire*. Wood and other flammable building materials can easily be ignited anywhere an exposed lightning channel comes in contact with (or passes through) them. It is most common for lightning to start a fire in the attic or roof of a house, as the channel usually has to pass through some of the structural material in the roof before it can reach a more conductive path such as wiring or pipes. When lightning current travels through wires, it will commonly burn them up - presenting a fire ignition hazard anywhere along the affected circuits.

- **Power surge damage**: If lightning chooses any of the home's electrical wiring as its primary or secondary path, the explosive surge can damage even non-electronic appliances that are connected. Even if most of the lightning current takes other paths to ground, the home's electrical system will experience enough of a surge to cause potentially significant damage to anything connected to it, electronics in particular.

- **Shock wave damage**: Another major source of damage from lightning is produced from the explosive shock wave. The shock waves that lightning create is what produces the thunder that we hear, and at close range, these waves can be destructive. Lightning can easily fracture concrete, brick, cinderblock and stone. Brick and stone chimneys are commonly damaged severely by lightning. Lightning's shock waves can blow out plaster walls, shatter glass, create trenches in soil and crack foundations.

Since we know the common paths lightning can follow in a house (wiring and pipes), the best thing to do is stay away from those paths as best as possible during a storm. Direct contact with them should be avoided. This includes taking a shower or bath, washing hands, doing dishes, typing on a computer, playing video games and using a wired phone, tool or appliance. Metal-framed windows should be avoided.

Direct skin contact with earth ground should also be avoided, as lightning current can travel through soil and across wet/damp concrete. Wear shoes if walking in a basement, garage or patio.

What to do if lightning strikes your house
If your home is hit directly by lightning, your immediate concern should be for any fires that may have been ignited. Again, the most common place for lightning-caused fires in a home is in the attic, but they can start anywhere the lightning has travelled. Some fires inside the walls and attic may not be immediately apparent and not easily

accessible. You should also watch for falling debris from damaged chimneys, shingles or walls.

You should also strongly consider contacting an electrician to have your home's electrical system inspected for any damage that might present a future fire hazard.

Can you get struck by lightning inside of a house?

While it is rare, yes, it is possible to receive a lightning injury inside a house. Burns and electric shock injuries can occur when someone is in direct contact with one of lightning's chosen paths to ground. The most common indoor lightning injuries involve a person talking on a corded phone or resting on/looking out of a metal-framed window. "Side flashes" (discussed above) that jump across rooms can also cause injuries, but are very rare.

How to protect sensitive electronics from a strike

It is nearly impossible to provide 100% protection to sensitive electronics from a direct lightning strike. The best thing to do is to unplug any high-value item you wish to protect during storms, as surge protectors and UPS units cannot provide direct-strike protection. Some could argue that the risk of a direct strike to any given house is too low to justify unplugging everything for every storm that passes overhead. It is wise to make sure your homeowner's or renters insurance covers lightning damage. Your insured expensive electronics can be replaced, after all. However, consider irreplaceable such as the data saved on your computer (photos, videos, work files, etc.). You can mitigate that risk by performing frequent offsite backups and/or storing data on an external hard drive that you can unplug when needed.

A professionally-installed and well-grounded lightning protection system will reduce or eliminate the fire and injury hazards if your house happens to be struck directly.

Get a Thunder Arrestor
Thunder storm delivers a huge amount of electrical charge. The work of a Thunder Arrestors to safely lead this current away from the building to the earth to avoid fire. Get one.

If protection fails or is absent, lightning that strikes the electrical system introduces thousands of kilovolts that may damage the transmission lines, and can also cause severe damage to transformers and other electrical or electronic devices. Lightning-produced extreme voltage spikes in incoming power lines can damage electrical home appliances or even leads to death. It can also use to protect electric fence

1.20 FIRE DUE TO CHARGING OF PHONE OR CHARGE PACK ON FLAMMABLE OR COMBUSTIBLE MATERIALS

Most people in this day and age have a smartphone. This means that the majority of us have access to our social media accounts, mobile games, news, e-mail, and more pretty much anywhere or anytime we need it.

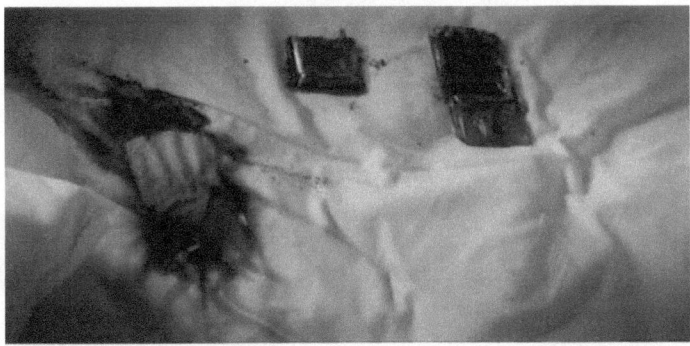

The problem with smartphones is that, because we use them so much, they usually run out of battery fairly rapidly. This means we all carry around extra chargers to bring to work, plug into our cars, and so on and so forth. All of this charging has, quite unexpectedly, led to the emergence of several charging-related myths. I am sure you are already familiar with a couple of them.

All that said, what is the truth about the Lithium-ion batteries powering our pocket computers?

What battery-related advice should you believe, and which should you forget about? Find out below.

1. Never Charge Your Phone Overnight

We have all likely heard this one before, and it probably emerged at a time when battery technology was nowhere near as advanced as it is today. The truth is that "leaving your phone plugged in overnight is okay to do."
Apparently, the technology regulating smartphone batteries has advanced to the point where it knows exactly when to stop feeding a charge into your device. In other words, there is no risk of you "overcharging" your phone and causing damage to the battery, as there are safeguards in place to prevent that from happening.
What you *do* need to worry about is overheating. So, if you are going to leave your phone charging overnight, make sure you place it in a relatively cool area. Also, remove any case

you may have put on it so that heat from the battery can escape in a timely fashion.

2. Let Your Phone Go To 0% before Charging
I don't know where this myth came from, but I've seen this repeated constantly. What makes this particularly **egregious** is that completely draining your battery before a charge actually causes it to become more unstable.
It is suggested instead that we keep our devices charged "between 50 and 80 percent." In other words, you should charge your phone intermittently throughout the day instead of waiting to perform a "deep charge" from 0 to 100 percent.

3. Any Charger, Even an Off-Brand Model, Will Work
While it may be tempting to try and save money by purchasing an off-brand charger for your phone, the damage it can do over time might make you think twice. The fact of the matter is that it is always best to use the charger that came with your device, even if you can find another cheaper model that still technically works.
Experts caution against off-brand chargers for simple reason: they are "not built with safety in mind." This means there is a far greater chance of these chargers causing a fire, or harming your battery, than there is with your phone's proper charger.

4. Turning off Your Phone Is Useless
While it might seem like an inconvenience to physically turn off our phones from time to time, experts suggest that we do exactly that. It has stated that "in order to maximize

battery life, you should [definitely] turn off your phone from time to time. "This does not mean that you have to always shut down your phone before bed, or do it on a daily basis. That would defeat the purpose of having an always-ready-to-use smartphone. You should, however, try and shut down or properly restart your device at least one a week, as this has been proven to conserve your device's battery life over time.

5. Do not Use Your Phone while it is Plugged In
As long as you are using the charger that came with your phone, or a certified replacement made by the same company, it is perfectly fine to use your phone while it is charging.
This myth does have a bit of a chilling origin, however. While it is safe to use your smartphone whilst charging it with its *proper* charger, it is not recommended to do so when using a *third-party* charger, as that may lead to the phone exploding, or worse, electrocuting the user.
While there is only a slim chance of that happening, you still should not risk it. Off-brand, third-party chargers might be cheap, as mentioned previously, but they do not work as efficiently with your phone's battery, meaning there's a much higher chance of it overheating and possibly injuring you or others during periods of extended use.

1.21 CLUTTER FIRE

Clutter is a common issue in offices and if regular cleaning and maintenance is not carried out, it will eventually increase the odds associated with a fire breaking out.
The office is full with combustible materials and fire hazards.

Ranging from a simple built up dust, grease, and overloaded refuse areas, to build up dirt and poorly ventilated areas can cause machinery and equipment to become overwhelmed with heat which can lead to a fire.
A regular cleaning regime and ensuring work areas are at an optimal state is integral for reducing potential fire risks.
Actively encouraging employees to keep their working environments as clean and tidy as possible will help to mitigate the risk of a fire breaking out hugely.

1.22 FIRE DUE TO HUMAN ERROR

Another major reason for fires inside the office is basic human error. This is typically because of incidents that were not intentional. A number of things can happen in a variety of ways including burning food in a staff area, spilling flammable liquids, improper use of machinery or equipment that overheats, and simple carelessness.

One method to prevent these types of incidents from escalating is to ensure there is plenty of suitable fire extinguishers located around the office or work area. It is also important that all employees are properly trained on how to use fire extinguishers and to regularly address and assess any potential risks in the office that could happen due to human error.

1.23 FIRE DUE TO NEGLIGENCE

Sadly, there are plenty of opportunities for fires to start at workplace due to negligence and lack of care. In an attempt to get a task done quicker, short cuts have been known to be taken where certain workers have chosen to ignore correct procedures to get the job done faster, which could indirectly cause a major fire, health and safety risk.

Examples: blocking of ventilation areas, stacking paper or card in a flammable area, misusing or improperly storing flammable or combustible materials, or overusing equipment or using equipment improperly. To help avoid these known issues, it is imperative to train your employees appropriately. Business owners should conduct reviews, retraining and assessments at regular intervals to ensure proper workmanship and reduce any risk of fires in the future.

1.24 FIRE FROM THE USE OF FIREWORKS.

Firework is a device containing gunpowder and other combustible chemicals which causes spectacular effects and explosions when ignited. It is used for display or in celebrations.

Fireworks are a class of low explosive pyrotechnic devices used for aesthetic and entertainment purposes.

Use of fireworks can be dangerous to you and to others and can cause serious public nuisance.

Fireworks can burn down houses, offices, damage entertainment venues such as nightclubs, frighten children and adults, cause bushfires, and terrify pets and wildlife.

Noise from fireworks can cause distress, especially as fireworks can sound like gunfire. The noise can also cause tinnitus and deafness, or aggravate a nervous condition.

People who suffer from asthma can experience discomfort and epileptics can experience seizures following fireworks displays.

Fireworks are often used to mark special events and holidays. However, they are not safe in the hands of consumers. Fireworks cause thousands of injuries each year. However, people can enjoy fireworks safely if they follow a few simple safety tips

- Be safe. If you want to see fireworks, go to a public show put on by experts.
- Keep a close eye on children at events where fireworks are used.
- Fireworks packaged in brown paper are made for professional displays – avoid buying.
- Always have an adult supervise fireworks activities, especially with sparklers.
- Back up to a safe distance immediately after lighting fireworks.
- Never point or throw fireworks at another person.
- Keep a bucket of water or a garden hose handy in case of fire and to douse used fireworks before discarding in trash.
- Never allow young children to play or ignite fireworks.
- Never carry fireworks in a pocket or shoot them off in metal or glass containers.
- Never try to re-light or pickup fireworks that have not ignited fully.
- Never place a part of your body directly over a firework device when lighting.
- Make sure fireworks are legal in your area before buying or using them.

CHAPTER TWO
FIRE EMERGENCY PREPAREDNESS

- Identifying the individuals responsible for various aspects of the plan (chain of command) so that, in an emergency, confusion will be minimized and employees will have no doubt about who has authority for making decisions;
- Identifying the method of communication that will be used to alert employees that an evacuation or some other action is required, as well as how employees can report emergencies (such as manual pull stations, public address systems, or telephones); and
- Identifying the evacuation routes from the building and locations where employees will gather.
- Install smoke detector at strategic places.
- Position fire extinguishers/ fire blankets at reach
- Train the household members what to do when the alarm is sounding and how to use portable fire extinguishers/blankets.

General Guidance for Fires and Related Emergencies

If you discover a fire or see/smell smoke, immediately follow these procedures.

- Notify the local fire department.
- Activate the building alarm (fire pull station); if not available or operational, verbally notify people in the building by shouting Fire! Fire!! Fire!!!
- Isolate the area by closing windows and doors and evacuate the building, if you can do it so safely.

- When evacuating the building, be sure to feel doors for heat before opening them to be sure there is no fire danger on the other side.
- If there is smoke in the air, stay low to the ground, especially your head, to reduce inhalation exposure. Keep on hand on the wall to prevent disorientation and crawl to the nearest exit.
- Go to your refuge area and await further instructions from emergency personnel.
- Shut down equipment in the immediate area, if possible.
- If possible and if you have received appropriate training, use a portable fire extinguisher to:
  o Assist oneself to evacuate;
  o Assist others to evacuate; and
  o Control a small fire.
- Do not collect personal or official items; leave the area of the fire immediately and walk briskly, do not run, to the exit and designated Mustering point or safe area.
- You should provide the fire/police teams with the details of the problem upon their arrival. Special hazard information you might know is essential for the safety of the emergency responders. You should not re-enter the building until directed to do so.
- If the fire alarms are ringing in your building, you must evacuate the building and stay out until notified to return. Move to your designated meeting location or move upwind from the building, staying clear of streets, driveways, sidewalks, and other access ways to the building. If you are a supervisor, try to account for your

employees, keep them together, and report any missing persons to the emergency personnel at the scene.

- Keep doorways, corridors and egress paths clear and unobstructed. Make sure that all electrical appliances and cords are in good condition.
- Do not tamper with any fire system equipment such as smoke detectors, pull stations or fire extinguishers.

If an individual is overexposed to smoke or chemical vapours, remove the person to an uncontaminated area and treat for shock.

Do not enter the area if you suspect that a life-threatening condition still exists (such as heavy smoke or toxic gases).

If you are Cardio Pulmonary Resuscitation (CPR) certified, follow standard CPR protocols. Get medical attention promptly.

If you or another person's clothing catches fire, extinguish the burning clothing by using the drop-and-roll technique, wrap the victim in a fire blanket, or douse the victim with cold water (use an emergency shower if it is immediately available). Carefully remove contaminated clothing; however, avoid further damage to the burned area. Cover injured person to prevent shock. Get medical attention promptly.

CHAPTER THREE

TECHNIQUES ON THE USE OF PORTABLE FIRE EXTINGUSIHER

While it is true that using a fire extinguisher is not rocket science there are a few basics you need to be aware of; and, probably are not. It is obvious that majority of human populace do not know how to use an extinguisher, even if they have one in their home, office or car. This is a dangerous knowledge gap. Fires double in size every 60 seconds, so you do not want to be fumbling around in an emergency situation, reading over the instruction manual and leaving a small flame on the stove grows into an inferno.

However, before you start to combat a fire;

- First, determine if the fire is one you can handle with your extinguisher. If it is bigger than what you can and the situation is unsafe, or the room is filled with smoke, get everyone out of the house.
- Secondly, if it up to what you can handle, position yourself with your back to an escape route, so you can make a quick getaway if necessary. Don't back yourself into a corner with just an extinguisher in hand.

The proper use of portable fire extinguishers can extinguish many fires while they are still small. These portable fire extinguishers are an important part of an overall fire safety program.

It is important to keep in mind that the successful use of portable fire extinguishers depends on the following:

- The portable fire extinguishers are properly located and in working order.
- The portable fire extinguishers are of the correct type.
- The fire is discovered while still small enough for use of the portable fire extinguishers to be effective.
- The fire is discovered by persons who are ready, willing, and able to use the portable fire extinguishers.
- That the extinguisher is properly updated i.e. had been exposed for monthly inspection by HSE staff.
- That the ring safety pin is not removed from its fixed position.
- That it is not depressurized – the point at the gauge is upright. If in doubt, call the safety officer in your area for confirmation.

Also, you should know that portable fire extinguishers are valuable for immediate use on small fires. They contain a limited amount of extinguishing material and need to be used properly so that this material is not wasted.

HOW TO USE PORTABLE FIRE EXTINGUISHER

- Carry the extinguisher from its location to the area of Fire outbreak.
- Pull the safety pin by removing it sharply (this also breaks the plastic seal). Test to ensure that the extinguisher is operable immediately after removing from mounting bracket.
- Always try to work in pairs for safety.
- Aim at the base of Fire, raising the extinguisher at an upright position.
- Squeeze the handles (lever) together to completely discharge the contents onto the seat or base of the fire.
- Sweep the Fire side to side. The Fire will be forced to extinguish.

- Operate extinguishers at their maximum effective distance (6 feet or 2 meters). Don't face the wind direction if it is at an open space.
- Avoid pointing Carbon Dioxide extinguisher applicators at people. If the extinguisher is accidentally operated, frostbite to the face and eyes may occur.

3.1 GENERAL FIRE SAFETY TIPS

- Consider replacing appliances before something goes wrong.
- Develop and practice an escape plan with your family.
- Do not use water to put out a grease fire.
- Store matches and lighters out of a child's reach.
- Avoid using water to extinguish a candle.
- Do not use candles during a power outage. Rely on flashlights instead.
- Clean your dryer lint screen regularly.
- Unplug toasters after each use.
- Inspect your dryer ducts each year.
- Do not store a toaster on your counter if you have pets.
- Contact an electrician if you are concerned about a wiring issue in your house.
- Select the right extension cord for the job.
- Avoid overloading electrical outlets.
- Avoid storing things in your microwave. Curious children can switch it on unknowing that things are store in it
- Never run an appliance, such as the dishwasher, washing machine, or dryer when you are not at home.
- Test your smoke detectors regularly. Consider replacing batteries when Daylight Savings Time begins and ends.

CHAPTER FOUR

CAUSES AND PREVENTION

OF

CAR FIRE

4.1 DESIGN FLAWS

A design flaw in a vehicle usually is not going to cause a car fire on its own, because there's no on/off switch for lighting a vehicle ablaze.

Design flaws, however, can make conditions really ripe for a fire, and sometimes even create conditions in which an eventual fire is inevitable. Usually, the manufacturers catch on to these situations before incidents become widespread. They issue recalls to get the dangerous cars off the street and fix the problems, because no carmaker wants to be known for combusting its customers. Like all automobile fires, a design flaw is only the first step leading to a blaze. Not all design flaws result in a fire, but any number of problems can make a fire a lot more likely.

4.2 POOR MAINTENANCE

Human error probably is not going to be the direct cause of a fire in your vehicle; after all, being lazy is not quite the same as striking a match and igniting a wick that goes into the gas tank. But if you are sloppy about maintenance, your car is going to be a lot more dangerous. In general, and the increased likelihood of a car fire is just part of the greater risks you are taking.

It is true, forgetting or neglecting to properly take care of your car can indirectly lead to a vicious fire. That is because if you let broken parts, leaky seals, or faulty wiring go without repairs, it will make your car a lot more hospitable to the conditions that cause a fire. An engine with a bad gasket is more likely to drip hazardous (and flammable) fluids. Frayed wiring is more likely to spark and make contact with flammable materials. Poor maintenance is actually one of the most common causes of car fires. Delaying the maintenance schedule, you are unknowingly increasing the risk of a car fire. Without maintenance, you cannot spot and repair broken parts, seals and wiring and your vehicle will slowly going in to the conditions of causing fire.

4.3 CAR CRASHES

Depending on the impact site, a car crash can even spark a car fire. Most vehicles' crumple zones are designed pretty well, so the sheet metal absorbs the force of a blow and protects internal, dangerous spots like the engine, the battery and even the gas tank. But really, there is not actually that much of a barrier there, so a hard enough hit is likely to cause fluid leaks and spillage, as well as heat and smoke.

And, as we know, high heat and spilled fluids create perfect conditions for a fire. Since it is hard for occupants of a crashed vehicle to see the extent of the damage while they are still inside, the threat of a fire might not be immediately apparent; however, it is always best to get away from a damaged car as soon as possible. Consider yourself lucky if you are not trapped inside a crashed vehicle; even if, it does go up in flames, at least you are a safe distance away.

4.4 ARSON

Arson (the criminal act of setting a fire); the question is why would anyone deliberately set a car on fire, anyway? It could be to cover up a theft, or to cover up the evidence of another crime. It could be old fashioned vandalism, too, wrecking something just for the sake of wrecking it.

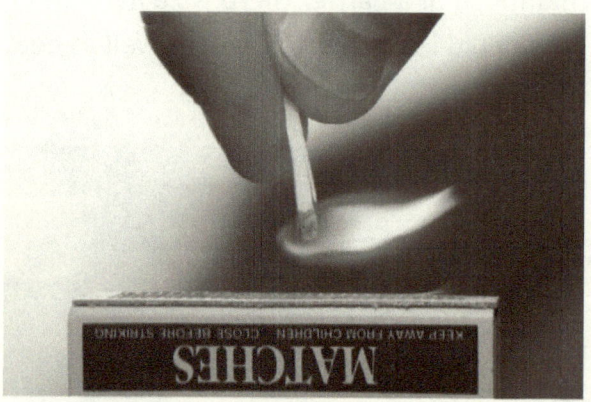

Or it could be insurance fraud. And there are probably several more reasons, but that is best left to the criminal masterminds. It is worth noting that it is pretty easy to set a car on fire; perhaps, doing it without being detected is a challenge, but actually igniting a car blaze is simple.

4.5 HYBRID AND ELECTRIC VEHICLE BATTERIES

Car battery is the part to look at to avoid car fires. Electric vehicle batteries was the problem with car mechanics from a long time ago. Even with the safest car in the world, the makers still need to perform many tests of vehicles caught fire during impact. In most of these problems, one of the major causes of car fires is the leaking coolant.

It interacts with the damaged batteries, which create sparking and blaze. This problem is amazingly troublesome, since there are always new potential risks with every new design.

4.6 OVERHEATING CATALYTIC CONVERTERS

Overheating catalytic converters are a fire risk that is often overlooked, but think about it. One of the consistently hottest parts of your car runs the entire length of the vehicle, the exhaust system. Catalytic converters usually overheat because they are working too hard to burn off more exhaust pollutants than they are designed to process.

In other words, if the car's engine is not operating efficiently (due to worn spark plugs or any number of other adverse conditions), it does not burn the fuel properly, and a lot of extra stuff ends up in the exhaust system. The catalytic converter then has to work extra hard to do its job, which makes it even hotter than usual. An overworked (or clogged) catalytic converter can easily go from its normal operating temperature range of about 1,200 to 1,600 degrees Fahrenheit (648.9 to 871.1 degrees Celsius) to up over 2,000 degrees Fahrenheit (1,093.3 degrees Celsius). This causes long-term damage not only to the catalytic converter itself, but to the car's surrounding parts. The car's designed to withstand the catalytic converter's normal temperatures, but it cannot consistently cope with temperatures several hundred degrees higher. If the catalytic converter gets hot enough, it could ignite the cabin insulation and carpeting right through the heat shields and metal floor pan.

4.7 OVERHEATING ENGINES

An engine that overheats and causes a car to catch on fire is an especially good example of how one problem can lead to another.

A car's engine probably would not overheat enough to simply burst into flames all on its own. But what can happen, is an engine can overheat and make the internal fluids, like oil and coolant, rise to dangerous temperatures and begin to spill out of their designated areas of circulation. When that happens, they drip, drizzle and spurt throughout the engine bay and onto the exhaust system, landing on other hot parts, where they can easily ignite and spread.

Generally, though, an overheating engine requires mechanical attention. There is often a leaky seal or gasket, or the radiator is not working properly, or any number of other things. If your car's engine is constantly overheating; well, that is not a symptom to ignore, very common, yet dangerous.

4.8 SPILLED FLUIDS

The average car or truck has a number of flammable and highly dangerous fluids under the hood: gasoline or diesel fuel, engine oil, transmission fluid, power steering fluid, brake fluid and even engine coolant.

All of those fluids are circulating when the car is on, and all of them can catch fire pretty easily if their lines, hoses or reservoirs take a hit. So even though one of the car's vital liquids is unlikely to start spewing or dripping out of nowhere; generally, something else has to go wrong first; the fact that all of these fluids are flammable to begin with is a problem in and of itself. Combined with another aggravating factor, like a car crash or a failed part, the result could be a fire. Though such a blaze is most likely to start in the engine bay, where all of these dangerous liquids are concentrated, keep in mind that some of them, like fuel and brake fluid, are moved along the entire length of the car.

4.9 ELECTRICAL SYSTEM FAILURES

Electrical system failures take the second spot on the list because they are the second most common cause of car fires. A typical car's standard battery is capable of causing plenty of trouble. The battery's charging cycles can cause explosive hydrogen gas to build up in the engine bay, and the electrical current the battery provides (along with faulty or loose wiring) can produce sparks that can quickly ignite a fluid drip or leaked vapours. The electrical system's hazards are not confined to the area under the hood, either.

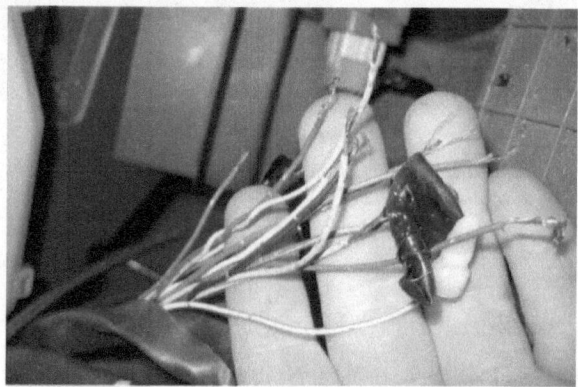

Electrical wiring runs throughout the entire car; through channels, into doors, under the carpet and through powered and heated seats, just to name a few places where a stray, unnoticed frayed wire could cause havoc. Electrical system failure can happen frequently so always be cautious. One more time, take your car to the car mechanics for schedule check to avoid the electrical system problem.

4.10 FUEL SYSTEM LEAKS

Leaks in the fuel system are the most common cause of vehicle fires, so that's why they take the top spot on the list. As we have already seen, any number of complicating factors can cause a fuel leak, but they are tricky because fuel leaks can also arise on their own and with very little warning.

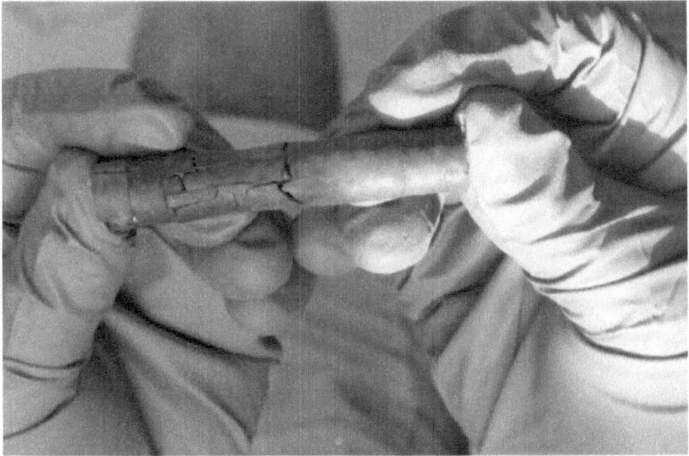

A fuel system leak is really dangerous. We have already discussed that a lot of a car's fluids have corrosive, poisonous and flammable properties, but gasoline is among the worst. Gasoline at a temperature of just 45 degrees Fahrenheit (7.2 degrees Celsius) or above can quickly catch fire from a simple spark. It happens all the time in a running car, after all, but it is contained by the engine. And gasoline that reaches 495 degrees Fahrenheit (247.2 degrees Celsius) will ignite by itself. It is easy to see how fuel dripping onto hot metal and plastic parts causes a fast-spreading fire. The best way to reduce chances of a fuel system fire is to make sure the car is properly maintained and to keep it out of the situations we have already described. And if you ever smell gas in or around your car, find and fix the leak immediately! Leaking fuel is very dangerous. Our recommendation is to make sure your vehicle is properly maintained and always check for leak before driving out.

4.11 TRANSPORT OF GAS CYLINDER OR PETROL IN A CAR BOOT

When transporting your gas cylinder inside your vehicle, ensure that it is firmly secured preferably upright position.

Put them in the boot or inside your vehicle in an upright position, and do not carry more than two bottles in the car at any one time.

Remove the cylinders from the car as soon as is practical. In other words, do not leave the cylinders in the vehicle unnecessarily.

Plastic gasoline containers are intended for brief periods of transportation, not for storage in confined spaces (car boot).

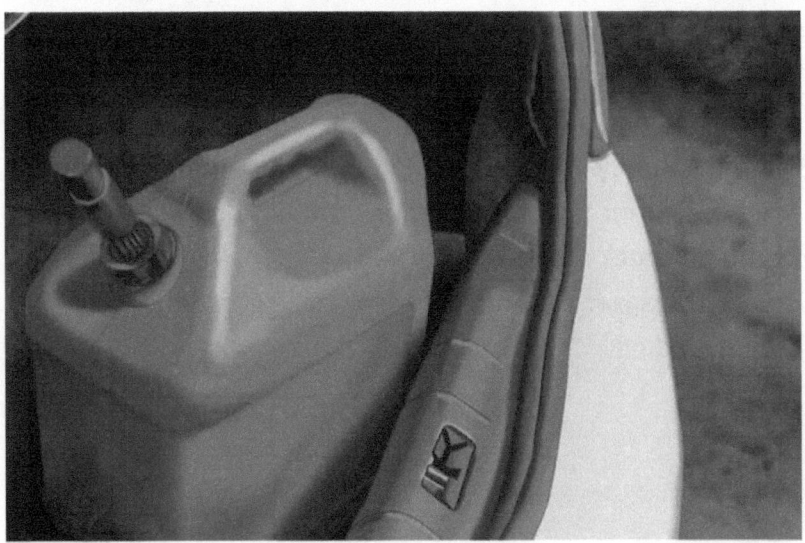

Transporting gas cylinder or plastic gasoline container inside car boot should be for a very short time. This is because hot atmospheric temperature, increases the temperature of your car boot or inside your car (if you are not using air conditioner) which in turn affects the temperature of gas or gasoline. The external temperature gradually affects the cylinder and its content and expands the gas molecules; and, as soon as it expands beyond the capacity of the cylinder, the cylinder will rupture into flames.

Also, when the temperature inside your car boot makes gasoline to reach its self-ignition temperature, it will rupture into flames.

Therefore, do not transport gas cylinder or plastic gasoline container for a long period inside your car boot. It should always be for a short time.

Furthermore, restrain all cylinders from moving during transportation.

Smoking is strictly forbidden when loading, transporting, and unloading any gas cylinder.

Close the gas cylinder valve and disconnect the regulator and hoses prior to transportation.

Regularly check for leaks from valves.

Do not transport a cylinder if a leak has been detected during loading.

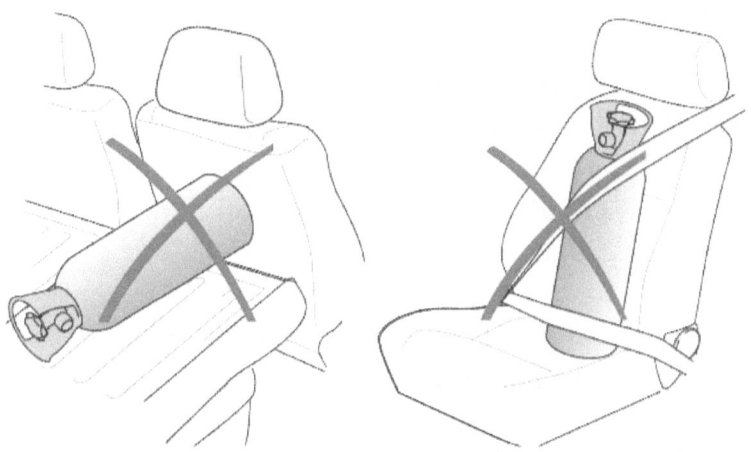

Transporting gas cylinders inside the driver or passenger's compartment as in pictures above is extremely dangerous.

When transporting a cylinder, if it cannot stand up securely in the boot of your car, put it inside the car and push it tightly behind the passenger seat to make sure that it does not fall over while you are driving as in figure below.

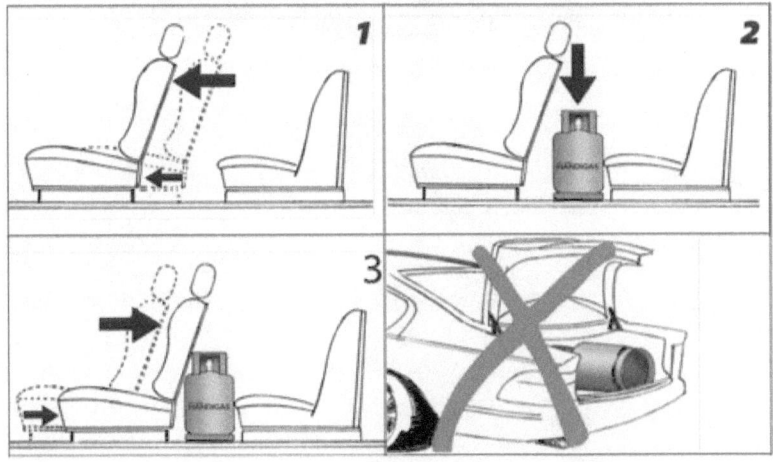

Carefully watch this method of carrying gas cylinder inside the vehicle because asphyxiation and fire could occur due to leakage. A leak in the vehicle with the glasses wind up (no ventilation) will form a gas cloud which is ready to be ignited in an explosive mode either from your handset or from your car electrical system {Refer to Amazon.com for *Awareness on Toxic, Flammable & Explosive Atmosphere*).

It seems good to transport gas cylinder between front and rear seats. Weigh your risk level and be wise.

CHAPTER FIVE

TIPS ON HOW TO PREVENT A CAR FIRE.

If you have ever driven past a vehicle on fire, you may worry that it could happen to you. Vehicle fires are not common, but they do destroy your property and can be life-threatening. Fortunately, you can prevent a car fire by maintaining your vehicle parts and wiring, staying safe while driving, using safe car habits, and observing warning signs. Others include:

1. **Car inspection.** Take your car to a maintenance shop regularly for inspection. A mechanic can give you a better idea about your vehicle's safety and any repairs that you need to get done. While it may seem like an added expense, it costs less to keep your car maintained than it does to replace a car that has broken down or, worse, caught fire.

2.   **Maintain your electrical systems.** Two-thirds of vehicle fires are caused by electrical system failures or malfunctions, so keeping your car's electrical system maintained is essential for reducing your fire risk. Make sure that your battery is in good condition and is properly hooked up, and check that none of your wiring is frayed or damaged.

3. **Check the fuel lines and tank.** Look for cracked fuel lines, bad fuel injectors, and fuel leaks. Bad fuel lines can lead to a flare-up, which can cause a vehicle fire.

If your tank is compromised, then it can leak fuel, increasing your risk of fire.

4. **Avoid storing flammable materials in your vehicle.** While you may occasionally need to transport items such as gas cans, lighter fluids, or propane gas, do not leave these items in your car on a regular basis because doing so can lead to fire. Only make short trips while carrying flammables, and immediately remove them from the vehicle once you have arrived at your destination.

5. **Avoid smoking in your car.** Smoking cigarettes can lead to a fire if you accidently leave a burning cigarette in your vehicle or if hot ash falls onto flammable material, such as paper or car seat. Your risk further increases if you carry a lighter to light the cigarettes.

6. **Keep your car clutter-free.** Throw out trash and avoid storing items in your car. Allowing these items to remain in your vehicle can create a fire hazard. Not only do extra items, especially crumpled paper, act as fuel if there is a spark, they also make it easier for you to accidentally leave a flammable item in your car.

7. **Carry a fire extinguisher and fire blanket.** A fire extinguisher and fire blanket can help you put out a fire. Purchase a fire extinguisher that is made for an automobile because the causes of a car fire are often

related to electrical issues or combustible fuel, which require a different type of extinguisher.

8. **Practice Defensive driving.** Vehicle fires can happen as part of a car accident, so avoid reckless or aggressive driving. While it can be frustrating to give up the right-of-way or to drive slowly, making choices that keep you safe can help you reduce your risk of fire.

9. **Watch for downed power lines.** Be careful when driving around after a storm has occurred because you can encounter dangers like downed power lines. If the downed power line still contains an electrical charge, it could ignite materials in or on your vehicle.

10. **Avoid driving if you have spilled oil.** During an oil change, you may accidentally spill oil on part of your vehicle. If this happens to you, wash away the oil before you try driving again. Having any amount of oil on your engine can lead to a vehicle fire.

11. **Choose a safe area to park.** The mechanical parts of your car can be hot, and parts of the system can ignite dry materials that they come in contact with. Make sure that you don't park around high grass or in an area where materials such as trash can come in contact with your undercarriage or catalytic converter. Instead, choose a flat area that is empty of debris, such as a driveway or the street.

12. **Watch for rapid changes in fuel or fluid levels.** If your car is leaking fluids, then you should see sudden, unexpected drops in your fuel or oil. Notice if you start needing to refuel more often or if the oil you just added to your vehicle is no longer showing on the dip stick. These can be signs that you have a leak that needs to be fixed immediately.

13. **Look for signs of an overheating engine.** An overheating engine can quickly lead to trouble. Even if it does not result in fire, it will likely result in a stalled vehicle. While an overheated engine can cause you a lot of headaches, it is not hard to spot the symptoms. Signs will include the temperature gauge warning light coming on, a smell of burning metal or rubber, a thumping or ticking sound, steam coming from under your hood, or a hood that is hot to the touch. You may also see that your coolant is low or leaking, and your car may not perform as well as it usually does.

14. **Notice blown fuses.** If your car has more than one blown fuse in a short period of time, then you need to get the engine checked. Blown fuses are a warning sign that something is wrong and that your car is at risk.

15. **Watch for cracked or loose wiring.** Cracked or loose wiring is a huge fire risk, so you need to get it

repaired immediately. Do not continue to drive a vehicle with damaged wiring. Check around your engine by looking over the wiring. Don't touch or remove anything. If you notice any wires sticking out from anywhere, get them checked out.

16. **Listen for loud sounds in your exhaust.** If you feel any clunking or cracking sounds in your exhaust system, then you may have a blockage or damage in your exhaust. Avoid having a build-up of gas or allowing a leak to continue by getting your exhaust system checked.

17. **Replace a missing oil or fuel cap.** A missing oil or fuel cap can be a hazard because oil and fuel are both flammable. Additionally, items can get into the system because of the missing cap, which can put your vehicle at risk.

18. **Check for broken or missing hoses.** While they are not that common, broken and missing hoses can compromise your vehicle or allow flammables to leak from your vehicle. Replace any hoses that you discover are damaged or missing. Look for leaks. If you notice that fluids in your car are dropping unexpectedly or that your air conditioning has stopped working, get your hoses checked. Glance under your hood to see if everything looks good and properly connected.

CHAPTER SIX

ACKNOWLEDGEMENTS

Uniquely, this book **"CAUSES & PREVENTION OF HOUSE, OFFICE AND CAR FIRE"** is dedicated to God Almighty for His infinite wisdom and power.

Thanks to my family, my beloved wife and children for their support and time. The production of this book would not have been possible without their sacrifices, self-denial, encouragements and prayers.

My cheers also go to the entire staff of Health, Safety and Environmental Department of Daewoo Nigeria Limited and Desicon Engineering Limited for their teamwork, support and opportunity to serve.

Many thanks to the editor of this book, Engr Ikechukwu Madu for his professional review and touch of this book. Engr Ikechukwu Madu is a senior Health, Safety and Environmental Professional of high repute; classically dedicated in saving lives.

I am eternally grateful to a brother and friend, Ugochukwu Cyriacus Okolie who out of his magnanimity and love fed me and provided shelter over my head. He taught me tough love and manners that so much helped me to succeed.

A very special thanks to Engr Humphrey Ezeifedi for allowing God to use him to provide a start-up career job for me even when it seems there was no hope for me.

Sincere thanks to Amazon Kindle Direct Publishing for shouldering the publishing of this book and making it enjoyable for readers.